The World of Nature

MONKEYS & APES

GALLERY BOOKS
An Imprint of W. H. Smith Publishers Inc.
112 Madison Avenue
New York City 10016

This edition first published in U.S.
in 1990 by Gallery Books,
an imprint of W.H. Smith Publishers, Inc.
112 Madison Avenue, New York, New York 10016

ISBN 0-8317-9592-1

Printed and bound in Spain

For rights information about the photographs in
this book please contact:

The Image Bank
111 Fifth Avenue, New York, NY 10003

Producer: Solomon M. Skolnick
Author: Alan Heatwole
Design Concept: Lesley Ehlers
Designer: Ann-Louise Lipman
Editor: Sara Colacurto
Production: Valerie Zars
Photo Researcher: Edward Douglas
Assistant Photo Researcher: Robert V. Hale
Editorial Assistant: Carol Raguso

Title page: **An adult lar gibbon perches comfortably in the crest of a tree while giving a morning howl. Both male and female gibbons howl in order to define their territory.**
Opposite: **A full-grown mountain gorilla stands in the lush vegetation of Rwanda in eastern Africa. The massive size of this species has not saved it from being hunted to the point of near-extinction.**

Far from the jungle that most chimpanzees call home, HAM, the Space Chimp whirls out into space at a speed of 5,800 miles per hour. In January of 1961, HAM made history and helped launch space exploration. Scientists selected the chimp because of the similarities between chimpanzees and humans. Such experimentation proved invaluable to the human astronauts yet to come.

Monkeys, apes, humans, and prosimians belong to the primate genus. Prosimians, or lower primates, are relatively small and extremely primitive. They look like dogs or raccoons, not other primates, and have fur with color patterns as opposed to hair. Monkeys and apes, however, resemble humans in terms of their appearance, skeletal structure, and intelligence. Next to humans, apes have the highest level of intelligence of any land animal.

Preceding page: **A young mountain gorilla reaches for leaves which it will use to pad its bed for an afternoon nap. Mountain gorillas are found in the eastern African countries of Uganda, Rwanda, and Tanzania, while lowland gorillas inhabit the countries of Cameroons, Gabon, and Zaire in western Africa.** *This page, above:* **Gorillas generally live on the ground, where they walk using both their arms and legs, but will climb into trees to build nests.** *Right:* **A family of mountain gorillas sits on a hillside. Social structure is very important to gorillas and they spend much of their time in family units, which are always dominated by a single, large male.**

Infant gorillas are rarely left alone, but as they mature they are often allowed to frolic in the trees. *Below:* A lowland gorilla is perfectly depicted in its natural habitat. *Opposite:* The head of an adult male gorilla appears massive due to the domed crest which attaches powerful neck and jaw muscles. They are very expressive creatures, using facial expressions and body language to indicate a whole range of emotions.

A lowland gorilla eats grass leaves in the Kahuzi Biega National Park. Gorillas are diurnal and strictly vegetarian.

There are more than 125 species of monkeys and four species of apes alive today. Gorillas, chimpanzees, orangutans, and gibbons are all apes. They inhabit only two continents, Africa and Asia. Monkeys outnumber apes and enjoy a much wider range, living on five continents. They abide in South and Central America, Southern Europe, Northern Africa, the Middle East, and throughout Southeast Asia.

All monkeys and apes have five digits on each of their hands and feet, with some variations. Spider, colobus, and Asian leaf monkeys, as well as the orangutan, have only a remnant of a thumb that hardly functions. Mandrills and baboons represent the typical primate, with very long thumbs. Most monkeys and apes depend on their opposable thumb for gripping and using tools. The advent of the opposable thumb ranks as high as any other single evolutionary advancement. Some zoologists argue that it has prompted the development of the primate brain.

Adult, male, mountain gorillas are also known as "silverbacks," because their coats are tinged with silver and grey. *Right:* Gorillas are peaceful creatures and will only attack if provoked or attacked.

Preceding page: Chimpanzees are very bright and resourceful. They are known to use twigs as tools, which they insert into termite mounds to extract the delectable insects. *This page, above:* Chimpanzees spend most of their time in small groups, although the strongest bonds are between the mother and her offspring, which remain with her for many years. *Right:* Chimpanzees resemble humans more than any other monkey or ape. They are extremely communicative, using gestures and vocal sounds to make a point.

Monkeys are divided into two categories by geographic distribution. New world monkeys live in Central and South America and old world monkeys range from Southern Europe and Africa through the Middle East to Asia. The two groups evolved separately because of the isolation of the continents; several different anatomical features distinguish the monkeys of either group.

All new world monkeys have tails, some of which are prehensile, or adapted for seizing or grasping. Monkeys use their prehensile tails as an extra hand. Often they can be seen hanging from a tree limb while enjoying a meal. Without exception, new world monkeys live in trees.

The nose also distinguishes the two groups. Noses of old world monkeys appear to be pointed, as the nostrils are set closely together. The broad, flat nose of the new world monkey differs sharply from that of the old world monkey: the nostrils are round and widely separated.

This page, top to bottom: **Two pygmy chimpanzees take turns grooming one another. This task helps to keep the animals clean and healthy as insects and parasites are removed from their fur, but it is also a valuable part of their social behavior. Although this pygmy chimpanzee is enjoying a daytime nap in the grass, it will most likely retire to a large tree-nest for its nighttime slumber.** *Opposite:* **A young chimpanzee shows the species, characteristic hairless face. Common chimpanzees have flesh colored faces, whereas the faces of pygmy chimps are black.**

Preceding page: The lar gibbon, a small ape native to Southeast Asia, is perfectly at home in its tree-top habitat. It is one of the most agile and fastest-moving of all the primates. *Above:* A lar gibbon rests on a branch after descending from a tree.

Gibbons live in family units consisting of a male, female, and several young. They guard their territory very carefully. *Below:* Lar gibbons are vegetarian and gather their food by hanging from one long arm and reaching for fruit with the other. *Opposite:* The long arms of the gibbon are perfectly suited for brachiation, or swinging through trees from arm to arm. Gibbons are less adept at walking than other species and spend most of their life off of the ground.

Old world monkeys have a greater range than new world monkeys, covering territory from the Rock of Gibraltar to the islands of Southeast Asia and to the tip of South Africa. Although most old world monkeys have tails, none of these are prehensile like those of the new world monkeys.

Only old world monkeys have cheek pouches – sacs of skin in the mouth near the molar teeth that store food for later consumption. A monkey fills its cheek pouch with chewed food and when hungry, ejects bits into its mouth to be swallowed. With a full pouch, a macaque sometimes resembles a baseball player with a mouthful of tobacco. As a monkey ages, its cheek pouches may stretch out of shape making empty sacs appear full. With facial muscles distended, the monkey may have to empty its pouch by hand. Cheek pouches are found on guenons, macaques, mangabeys, and baboons. Male monkeys occasionally remove bits of food from the cheek pouches of females, possibly to display dominance.

Preceding page: **A lar gibbon mother and her infant eat leaves on the forest floor.** *This page, above:* **Kloss gibbons are also found in Southeast Asia.** *Below:* **Their coloring is much darker than the lar gibbon, although their method of transportation and food gathering is the same.**

Animals in captivity are much easier to study than animals in the wild. However, animals do not always behave normally when confined. Fortunately, over the past 30 years there have been several lengthy field studies made by such notable researchers as Jane Goodall and Dian Fossey. Such studies have yielded a wealth of information and helped highlight the need for species protection as well as habitat conservation. Loss of their jungle habitat is the leading cause of species extinction.

Most monkeys live in the jungle of South America and Africa. The floors of these jungles are made up of a thick underbrush of shrubs and bushes. Small trees also form a zone 20 to 50 feet in height. This is a closed canopy in which the limbs and branches of the trees overlap. As trees grow out of this interwoven layer their leaves and branches spread slightly, making contact but no longer overlapping. These trees usually grow up to 100 feet. The tallest trees of the jungle unfold into an open canopy. These trees grow far apart from one another and may reach a height of 150 feet.

An adult orangutan of Borneo examines his hand. Unlike other apes, these highly intelligent primates spend most of their life in solitude. They are one of the most endangered of all mammals, with less than 3,000 left in the wild. *Following page:* A rare glimpse of an orangutan in the wild. Apes occasionally eat flowers, along with their customary diet of fruits, leaves, shoots, and insects.

Preceding page: An infant orangutan clutches its mother. Mothers will often nurse their young for more than two years.
This page: While still holding tight, the infant nibbles at a leaf. When it is finally time to separate from its mother the young orangutan will characteristically wander off to live the remainder of its life alone, except for brief mating periods. *Opposite:* A young orangutan in the Sepilok Forest Preserve in Borneo scratches its back. Though they may stand upright occasionally, orangutans find walking on two feet extremely difficult because of their body structure.

This layering of foliage presents different habitats, and monkeys have evolved with characteristics specially suited to each layer. The monkeys living in the open canopy atop the jungle swing and can jump greater distances than the slower moving species of the lower zones. The interlaced trees of the closed canopy provide an intricate set of paths high above the ground.

Life in the trees means eating, sleeping, and moving high above the forest floor. Monkeys and apes are adept at using their arms – and often their tails – to swing from branch to branch. Zoologists call this brachiation. The motions of the arms can be likened to the sequence of human walking – left, right, left, right. When pivoting about a tree limb, the body of a brachiating ape swings like a pendulum.

This page, top to bottom: **Found in Madagascar, ring-tailed lemurs are known to live on the ground, although they will climb trees in search of fruit, leaves, and other food. The red ruffed lemur of Madagascar displays both nocturnal and diurnal behavior. Brown lemurs are also found in Madagascar.** *Opposite:* **Lemurs are unique to the rain forest of Madagascar. They are extremely rare and among the most threatened mammals in the world.**

Preceding page: Philippine tarsiers have huge eyes which dilate at night, when they hunt for prey. During the day they remain in the trees. *This page:* A young hamadryas baboon sits and eats the leaves it has gathered. Baboons have a hardened pad for sitting, as they spend most of their time on the ground.

Apes are classified as modified brachiators because of their anatomy – long arms, hook-like hands, and wide chests. Only occasionally do they show the pure form of arm-swinging locomotion. In the case of gorillas they probably never move in this manner except when very young. The chimpanzee and the gorilla are only partly arboreal, or tree-living. Chimpanzees sleep in trees, but spend about half of their day on the ground. Gorillas live most of their lives on the ground.

Humans are the only primates that can truly walk upright. Monkeys and apes are quadrupedal – they use both arms and legs together when walking. The anatomy of apes does not lend itself to bipedal walking. Although apes occasionally walk on two legs, their gait is slow and clumsy and they never walk long distances standing erect. When walking on all fours, they "knuckle-walk." As they knuckle-walk, the body remains semi-upright, for their arms are considerably longer than their legs.

The gait of orangutans is quite different from that of the other great apes. It is a form of knuckle-walking, but they put their weight on their bunched fists rather than their knuckles.

Hamadryas baboons adhere to a strict hierarchial order. Infants will go everywhere with their mothers, riding on their backs. *Following page:* Baboons are large but slender monkeys. If attacked, males will engage in fierce combat to protect the group.

Preceding page: A chacma baboon nurses her young in the Kruger National Park of South Africa. *This page:* Baboons are very agile rock climbers, moving about in large groups in search of food. *Left:* A couple of guinea baboons share the task of grooming. *Opposite:* Both the male and female gelada baboons have red skin patches on their chests. Males will expand their chests to frighten adversaries.

An adult gelada baboon is at home in the cliffs of northern Africa. *Left:* The mandrill is a very colorful baboon, distinguished by a red and blue face and a white chest. *Opposite:* Female mandrills lack the intense red and blue coloring of the face, present in males.

They spend only 15 percent of their time on the ground and are the least capable bipedal walkers – it is doubtful that they ever walk on two feet in the wild. The weight of their body rests on the outer borders of their feet, whereas humans shift their weight from the heel to the ball of the foot, and then to the big toe. At one time, zookeepers tried to teach orangutans to walk on two legs but the result was an awkward stride.

When climbing in trees orangutans use both arms and legs to suspend their body. They move very slowly, most prudently, and suspend themselves from their feet frequently, more often than either chimpanzees or gorillas. It is common for orangutans to fall considerable distances in the trees. This is known because of the many healed fractures discovered upon examination after an animal's death.

Gibbons, too, walk short distances on two feet but seem unsure of what to do with their long arms, sometimes holding them over their heads. Likewise, many monkeys, such as the patas and the vervets, can stand upright and some, like macaques and spider monkeys, can walk bipedally for very short distances.

Preceding page: **Besides the blue and red central area of the face, adult male mandrills have white whiskers, an orangish goatee, and large canine teeth.** *This page,* **top to bottom: Japanese macaques are the only primates, other than humans, that can survive cold, snowy winters. An adult Japanese macaque grooms a young macaque in a field of snow. A Japanese macaque infant's playing seems defiant as an adult monkey grooms itself.**

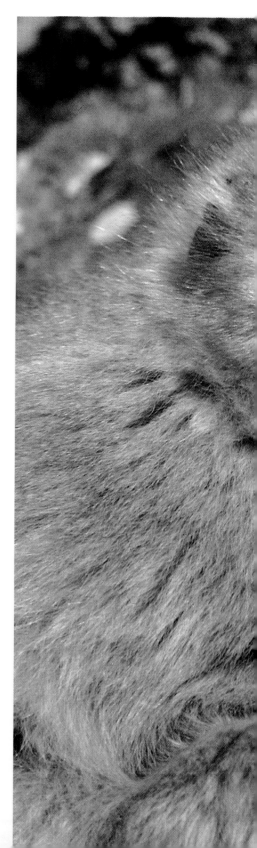

Preceding page: Huddled next to a rock, two Japanese macaques brave a snowstorm. These monkeys inhabit the mountains of Japan and mainland Asia and are unencumbered by treacherous conditions. *Below:* A family of Japanese macaques clings together for warmth. *Following page:* Japanese macaques can subsist on tree bark through harsh winters, when leaves and shrubs are scarce.

Preceding page: An adult Japanese macaque stares curiously upward. *This page:* Long-tailed macaques rest in a tree in Singapore. All macaques are very capable swimmers. *Opposite:* Macaques have sharp teeth and powerful jaws, which enable them to crack open most fruits and nuts.

Most monkeys and apes eat a variety of foods, although a few have specialized diets. Gorillas feed almost exclusively on ground plants in the wild, but often eat fruit and minced beef in captivity. Arboreal monkeys of Africa, Asia, and South America eat fruit and leaves along with bird eggs, grubs, insects, tree-frogs, and tree sap. In the wild, baboons eat fruit, leaves, and even animal flesh. Colobus monkeys of Africa and leaf monkeys of Asia rely mainly on leaves and shoots. Ground-living species, such as gelada monkeys, have a staple diet of young grass shoots, supplemented with leaves and seeds.

Mammals groom their fur just as birds preen their feathers. However, unlike a bird, which cares solely for itself, monkeys and apes groom each other. Grooming helps keep the fur clean and the animal healthy. It also serves as a valuable social bond, uniting mothers and their offspring in regular grooming parties. Young animals play with adults in a relaxed atmosphere, which strengthens family ties. The cohesion of the group or family unit is fortified as grooming communicates feelings of well-being among all its members.

Macaques live in troops of 10 to 50 individuals. This group of long-tailed macaques, in Bali, Indonesia, huddles together, with several drifting off to sleep.

Grooming is also essential in the social structure of macaques. Here a celebes crested macaque grooms a dominant male. *Below:* Like other species, infant macaques often ride on the backs of their mothers. *Opposite:* A close-up of an adult male pigtail macaque reveals a dark patch of fur on the top of its head.

Preceding page: A rhesus macaque rests on a tree limb. Macaques have a wider range of habitat than any other primate except man. They feed on seeds, fruits, berries, insects, and other small creatures. *This page, right:* A rhesus macaque meticulously searches the arm of a familial monkey for insects or parasites. Hard to reach places such as the elbow and the head undergo particular scrutiny during grooming. *Below:* Rhesus macaques are territorial, but physical battles are very rare.

A monkey hoping to be groomed will approach another monkey and present part of its body. The other monkey usually responds by parting the fur to examine the underlying skin. The groomer picks off flakes of dry skin, dirt, and parasites. He carefully cleans each area of skin and pays particular attention to the face, shoulders, back, and to scratches or wounds. Then the roles are reversed and another session starts. Groomers are persistent and can remain absorbed in the task for hours. The relaxed expression of the monkey being groomed explains why these sessions can last so long.

Like humans, monkeys and apes are usually quite gregarious. Three groups prevail and most species of monkeys and apes live in one. Zoologists distinguish these groups by the number of adult males present and the degree of competition among them. Monkeys and apes are most commonly found in family groups, groups dominated by a single male, or multi-male groups.

The family group resembles the human nuclear family. A mated pair bonds for at least the duration of their offsprings' development. The adult male helps the mother care for the young. Families may congregate in larger groups, or herds, but the individual family unit remains intact as bonding among females of the herd does not occur.

Another type of group consists of a single dominant male, several adult females, and their young. The adult male must maintain his authority within the group by confronting any outside males searching for a group. Eventually, the group will be taken over by a competing male. There is bonding among the females of the group and between the females and the young of the group. This bonding is crucial in defending the young against adult males of the same species. Often several adult females charge an adult male to protect a young monkey. It is presumed that adult males kill the infants of other males in order to enhance the survival rate of their own offspring. The accomplished adult male will produce several offspring during his reign and watches them age past vulnerability. He does not need to stay with this particular group for his own survival and when ousted will probably find a place in another group.

The black-faced vervet is native to Africa, although it has been introduced to the Caribbean. *Left:* Much more adept at moving through the trees, black-faced vervets usually walk on all fours when on the ground. *Opposite:* A male hamadryas baboon bares his long, sharp, threatening canines in a typical show of dominance.

Multi-male groups consist of several adult males, several adult females, and their offspring. The adults are promiscuous. During the oestrus cycle adult females mate freely and frequently, making parentage impossible to ascertain. All members of the group, including adolescents, share the responsibility of rearing the offspring.

The distinctions of these groups often overlap. For example, single-male dominated groups of geladas and hamadryas baboons form herds of 100 or more at night. They sleep in the steep cliffs of Northern Africa and find safety in numbers. Another exception to single-male groups can be seen with several species, such as the orangutan. These groups consist of several females living in overlapping home ranges. A single adult male regularly visits several ranges.

Chimpanzees live in a loose form of the multi-male group. These groups of up to 80 chimps break into small bands, which constantly change as they meet, mingle, and separate during their daily forage. Chimpanzees often greet one another with affectionate gestures. They will embrace, hug, kiss, hold hands, touch, or pat each other, depending on the occasion and their friendship.

The langur monkey is considered a sacred animal in India. *Left:* Patas monkeys live in the driest regions of Africa. They are very social, using gestures and facial expressions to communicate. *Opposite:* The Vietnam War, with its unprecedented use of defoliants, took a heavy toll on the world's population of douc langurs.

A black and white colobus monkey pauses for a moment in the treetops of central Africa. *Below:* White-faced capuchins live in large groups in the rain forests of southern-Central America.

A group of squirrel monkeys plays in the upper canopy of the Amazon Rain Forest of Brazil. Their prehensile tails enhance their ability to swing through the trees. *Below:* Squirrel monkeys spend most of their lives high above the jungle floor.

Howler monkeys are named after their distinctive nighttime howl, which can be heard up to two-miles away. *Left:* Black howler monkeys live in the rain forests of Belize.

Spider monkeys of South America will sometimes forage for food on the ground. *Right:* As tropical forests are depleted, these already endangered animals might be completely wiped out.

This red uakari, native to the rain forests of South America, clings to the limb of a tree. Zoologists can determine the health of these monkeys by looking at the face: a healthy uakari has a bright, red face. *Below:* Squirrel monkeys rarely leave their arboreal habitat, where they feed on berries, insects, frogs, and other small animals they find in the trees.

When frightened, a chimpanzee will touch another group member for reassurance.

One of the most endangered species of all mammals is the orangutan. There are now less than 3,000 in the wild. The rampant deforestation of the earth – about 150 acres per minute – is rapidly changing the lifestyles of monkeys and apes. The lack of trees and the resultant change in food supply have forced several species to spend much more time on the ground. For all monkeys and apes, the ability to adapt to changing environments is crucial if the species is to avoid extinction. Global conservation and wildlife protection is crucial to their survival, as well as to our own.

This page, top to bottom: **Common marmosets are new world monkeys, all of which have tails. The golden lion tamarin is native to Brazil and is characterized by its thick, lionlike mane. Woolly monkeys, native to the Upper Amazon Basin of South America, use their prehensile tails as a fifth limb, to swing through the trees with great agility.**

Proboscis monkeys have very large noses. As males age, their noses can become so long that they must be moved with the hand in order for the animal to eat. Male proboscis monkeys give loud honks which sound particularly resonant as the snout straightens out.

Index of Photography

TIB indicates The Image Bank